HOW PEOPLE LIVE

ON THE COAST

Nancy Dickmann

BROWN BEAR BOOKS

Published by Brown Bear Books Ltd
4877 N. Circulo Bujia, Tucson, AZ 85718, USA
and
Studio G14, Regent Studios, 1 Thane Villas, London N7 7PH, UK

Text: Nancy Dickmann
Design Manager: Keith Davis
Children's Publisher: Anne O'Daly

Library of Congress Cataloging-in-Publication Data
Title: On the coast / Nancy Dickmann.
Description: Tuscon, AZ : Brown Bear Books Ltd., [2025] | Series: Fast track: How people live | Includes bibliographical references and index. | Audience: Ages 5-7 years | Audience: Grades K-1 | Summary: "How people live by coasts all around the world"– Provided by publisher.
Identifiers: LCCN 2023052090 (print) | LCCN 2023052091 (ebook) | ISBN 9781781219690 (library binding) | ISBN 9781781219751 (paperback) | ISBN 9781781219812 (ebook)
Subjects: LCSH: Coasts–Juvenile literature. | Seashore ecology–Juvenile literature. | Seashore biology–Juvenile literature. | Seaside architecture–Juvenile literature. | Boats and boating–Juvenile literature. | Seashore–Recreational use–Juvenile literature.
Classification: LCC GB453 .D556 2025 (print) | LCC GB453 (ebook) | DDC 307.0914/6–dc23/eng/20240105
LC record available at https://lccn.loc.gov/2023052090
LC ebook record available at https://lccn.loc.gov/2023052091

The photographs in this book are used by permission and through the courtesy of:
Cover: Shutterstock: Andrey Gudkov. Interior: iStock: Daniel Bendjy 20t; Shutterstock: Simon Annable 7, Barmalin 10, Curiosity 1bk, 2-3, 20-21bk, 22-23bk, 24, Louis-Michel Desert 23, Dmitry Tkachenko Photography 21t, Epic Stock Media 18, GizemG 4, Good Studio 8-9bk, 16-17bk, Riekelt Hakvoort 6, Simone Hogan 13, IgorZh 8, Laborant 9, Pierre Laborde 21b, Lake View images 12, Carlos Pereira M 14, MaDosa 19, mubus7 5, Nyky89 11, Oceloti 4-5bk, 12-13bk, Viktor Petruk 15, Porcupen 20-21c, Romeona 21cr, Victoria Sergeeva 24br, Boris Stroujko 16, SurfsUp 1c, the8monkey 6-7bk, 14-15bk, Theodore Trimmer 17, Maarten Zeehandelaar 20b, Alyona Zhitnaya 10-11bk, 18-19bk.
t-top, b-bottom, l-left, r-right, c-center, bk-background
All other artwork and photography © Brown Bear Books.

Brown Bear Books has made every attempt to contact the copyright holders. If you have any information about omissions please contact: licensing@brownbearbooks.co.uk.

Words in **bold** appear in the Words to Know on page 23.

Manufactured in the United States of America
CPSIA compliance information: Batch#AG/5659

Contents

At the Ocean's Edge

Look at a photo of Earth from space.

You can see land. You can see huge oceans.

Coasts are where the land meets the sea.

People often visit coasts.

Satellites in space take pictures. Some show the shapes of coasts.

WOW!

Coasts can change shape. Waves crash against the shore. They wear away rock.

Some coasts have flat, sandy beaches. Others are rough and rocky. A few have steep **cliffs**. Coasts are home to many plants and animals. People live there too!

Weather and Climate

The coasts near the **poles** are cold.

Coasts near the **Equator** are much warmer.

Huge storms sometimes build up at sea.

They often hit the warm coasts.

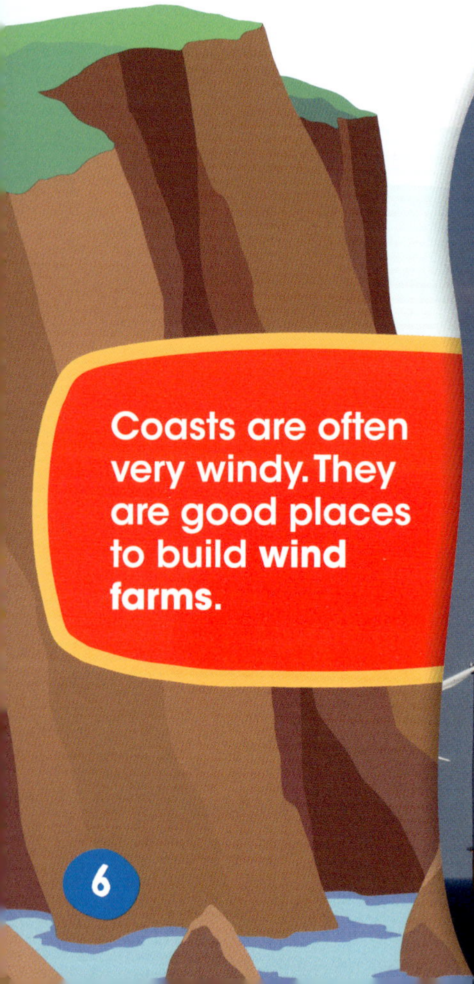

Coasts are often very windy. They are good places to build wind farms.

The sea level rises and falls twice a day. These are called **tides**. At high tide, the sea comes in. It covers the shore. At low tide, the sea goes out.

Coastal Homes

People like to live by the coast.
But the houses must not be too close
to the sea. They must be strong.
They have to stand up to storms.

Some coastal areas are steep. Houses are built into the slope.

WOW!

Wind and waves wear away coasts. The land washes into the sea. Homes built on this land are swept away.

Some people build their homes on stilts.
They live high above the water.
Even in a flood, their house will stay dry.

Food and Drink

Land near coasts is often salty. Many **crops** don't grow well there. But other plants do. People collect seaweed. It's tasty and healthy, too!

Samphire grows by the coast. It has a salty taste.

There is plenty of seafood along the coast. People catch fish. They find seafood on the shore. Mussels cling to underwater rocks. People gather them when the tide goes out.

Clothes

People on coasts once made their own clothes. They used materials found nearby. Some people collected sea grasses. They turned them into cloth. Others made cloth from palm leaves.

On some islands, people made cloth from tree bark. They still wear it on special occasions.

Today, people on coasts dress for the weather. A jacket protects them from wind and rain. A wetsuit keeps them warm in the water. They wear hats on sunny days.

WOW!

Some wetsuits are made from plastic bottles. This helps stop plastic pollution.

Jobs

Some coasts have beaches. They are great for vacations. Many local people have jobs helping **tourists**. They work in hotels and restaurants. They may have a different job in the winter.

Lifeguards watch over swimmers. They keep them safe in the water.

Some people go to sea to catch fish. They go in boats called trawlers. The boats have big nets. The nets scoop up fish.

Getting Around

Many people drive to the coast. But some coastal villages are cut off. No roads go there. People use boats instead of cars. They sail along the coast.

Boats are kept in a harbor. It protects them from the waves.

The Pacific has many small islands.
People travel between them by boat.
Some use canoes. These boats are light.
People use paddles to move them.

Games and Sports

Tourists visit coasts to enjoy the sea. They go bodyboarding in shallow water. People surf the waves. They sail boats. Some people like to go fishing.

Jet skis are very fast! They skim along the water's surface.

There is fun on land, too. Many coasts are beautiful places to walk. The views are amazing! Some beaches have sand contests. People build amazing structures with sand.

WOW!

There are canoe races in Hawaii. Teams paddle for 30 miles (48 km). They go from one island to another.

Where in the World?

There are coasts all around the world. Here are some of them.

Hawaii is made of many small islands. Its surfing beaches are famous.

Volleyball is popular on the beach at Rio.

Norway's coast is rugged and rocky.

There are many fishing villages along Thailand's coast.

Senegal's sandy beaches are a great place to play soccer with friends.

Activities

Look on the map and find the coast that is closest to you. How far away is it? Use the map scale to figure it out. How would you get there?

Choose a coastal location and look up the tide tables. When is the next high tide? How much do the tides go up and down?

Did you know that when the tide goes out, it leaves sea creatures cut off in rock pools on the beach? Research the animals that make their homes in these little pools.

Find Out More

Websites

dkfindout.com/us/earth/coasts/

kids.britannica.com/kids/article/coast/476241

scijinks.gov/tides/

Books

On the Coast Claudia Martin, QEB Publishing 2019

The Seashore: Explore Nature Lisa Regan, Wayland 2019

Waves Meg Gaertner, Cody Koala 2020

Words to Know

cliff a very steep rocky face

crops plants that people grow to eat

Equator an imaginary line around the middle of the Earth, where the weather is usually warm

flood an overflow of water onto dry land

harbor an area on the coast that is protected from waves, where boats can be kept safe

poles the areas at the most northern and southern points on Earth, where it is very cold all year round

satellite a spacecraft that travels in orbit around Earth

tides the daily rise and fall of the sea level along the coast

tourist a person who travels to a place on vacation to see the sights

wind farm a place with many windmills that turn wind into electricity

Index